ANIMAL GROUPS

A CRASH OF RHINOS

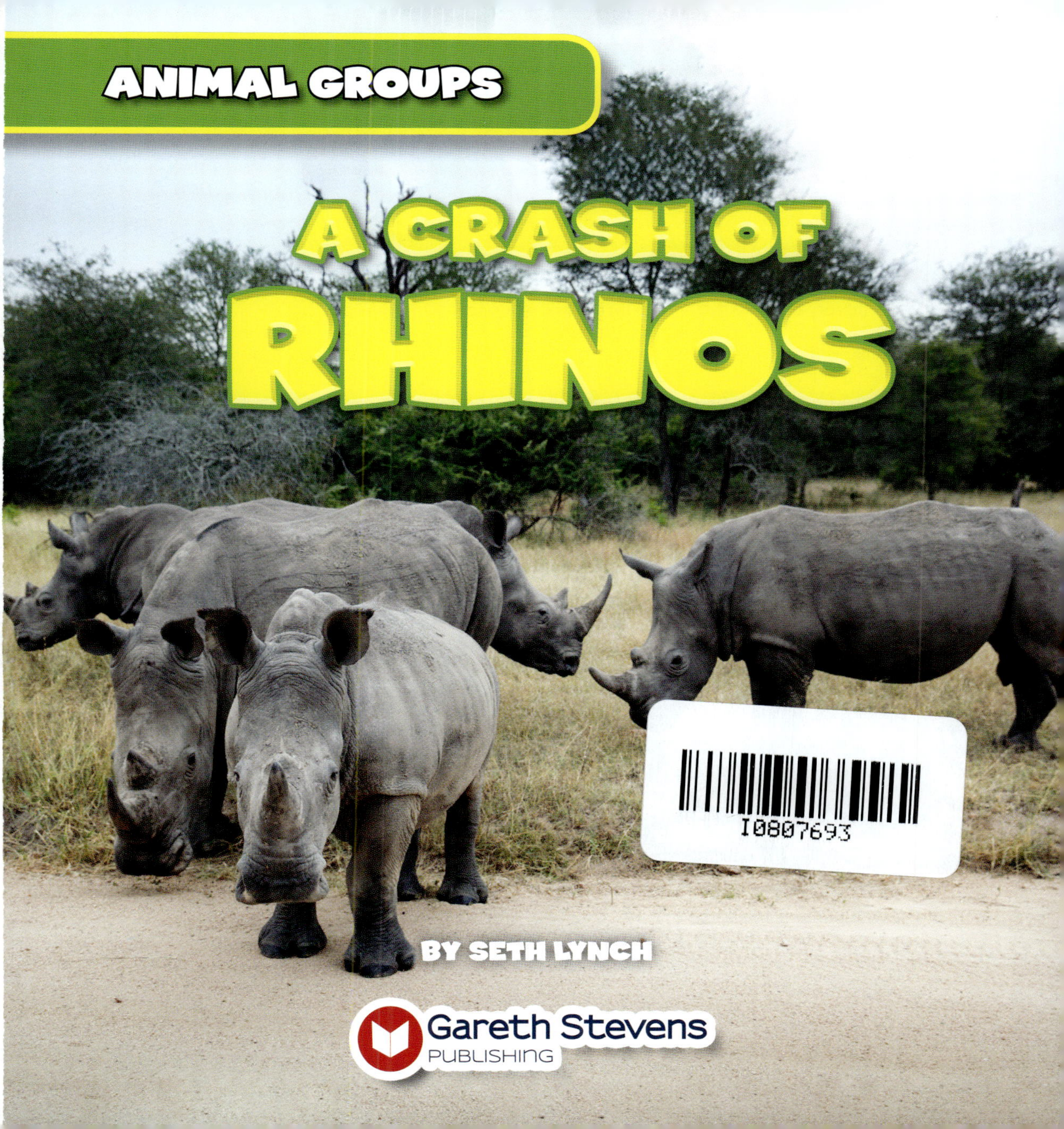

BY SETH LYNCH

Gareth Stevens
PUBLISHING

Please visit our website, www.garethstevens.com. For a free color catalog of all our high-quality books, call toll free 1-800-542-2595 or fax 1-877-542-2596.

Library of Congress Cataloging-in-Publication Data

Names: Lynch, Seth, author.
Title: A crash of rhinos / Seth Lynch.
Description: New York : Gareth Stevens Publishing, [2022] | Series: Animal groups | Includes index. | Audience: Ages 6 | Audience: Grades K-1
Identifiers: LCCN 2020028906 (print) | LCCN 2020028907 (ebook) | ISBN 9781538263518 (library binding) | ISBN 9781538263495 (paperback) | ISBN 9781538263501 (set) | ISBN 9781538263525 (ebook)
Subjects: LCSH: Rhinoceroses–Juvenile literature. | Rhinoceroses–Behavior–Juvenile literature.
Classification: LCC QL737.U63 L96 2022 (print) | LCC QL737.U63 (ebook) | DDC 599.66/8–dc23
LC record available at https://lccn.loc.gov/2020028906
LC ebook record available at https://lccn.loc.gov/2020028907

Published in 2022 by
Gareth Stevens Publishing
111 East 14th Street, Suite 349
New York, NY 10003

Designer: Tanya Dellaccio
Editor: Kristen Nelson

Photo credits: Cover Paul Souders/Stone/Getty Images; p. 5 RudiErnst/Shutterstock.com; p. 7 PhotocechCZ/Shutterstock.com; p. 9 Thorsten Spoerlein/Shutterstock.com; pp. 11, 15 Rudi Hulshof/Shutterstock.com; p. 13 costas anton dumitrescu/Shutterstock.com; p. 17 Sue Berry/Shutterstock.com; p. 19 JoseteMartin/Shutterstock.com; p. 21 Cathy Withers-Clarke/Shutterstock.com.

Printed in the United States of America

CPSIA compliance information: Batch #CSGS22: For further information contact Gareth Stevens, New York, New York at 1-800-542-2595.

CONTENTS

Boldface words appear in the glossary.

Rhino Rally

Have you ever seen rhinoceroses at the zoo? These plant eaters are some of the biggest land **mammals** on Earth! They range in color from gray to brown and have skin folds that make it look like they're wearing **armor**.

Kinds of Rhinos

There are five different kinds of rhinos. White rhinos and black rhinos live in the southern and eastern parts of Africa. They live on grasslands. White and black rhinos both have two horns. They have some of the longest horns of any kind of rhino.

WHITE RHINO

Sumatran rhinos also have two horns. They are the only rhinos covered with hair! Javan rhinos and greater one-horned rhinos only have one horn. They live in Asia. Sumatran rhinos do too. These three kinds live in **swamps** and rain forests.

GREATER ONE-HORNED RHINO

Crash!

A group of rhinos is sometimes called a crash! Adult rhinos weigh over 1,000 pounds (450 kg) and as much as about 5,000 pounds (2,270 kg). So, if they run into anything, it would really cause a crash!

In the Group

Rhinos don't live in big groups. Two to three males live in one **territory**. One is the dominant male, or leader. They mark the borders of their territory, but other males mostly keep to themselves.

Female rhinos have larger **ranges**. Some kinds live in small groups of females and young. They may move through many males' ranges. When it's time to **mate**, a female goes into a male rhino's territory to meet the male.

Rhino Calves

Mother rhinos commonly have one baby, or calf, about 16 months after mating. It will be about 100 pounds (45 kg) at birth! The mother keeps her calf safe from predators. Calves stay with their mothers for up to four years.

Talking to Each Other

Rhinos **communicate** with one another through many kinds of sounds. Some humans can't hear but other rhinos can! Rhinos also may roar if they get in a fight. They snort to show anger or worry. They may even scream to show they are afraid.

Rhinos in Danger

Crashes of rhinos are becoming much less common in the wild. In fact, black, Javan, and Sumatran rhinos are in great danger of dying out. Zoos, nature **preserves**, and **conservation** groups are trying to save these huge, amazing animals!

GLOSSARY

armor: a thick covering worn to keep someone safe from harm

communicate: to share thoughts or feelings by sound or movement

conservation: having to do with the care of the natural world

mammal: a warm-blooded animal that has a backbone and hair, breathes air, and feeds milk to its young

mate: to come together to produce babies

preserve: a safe place set aside for animals and plants

range: the area in which an animal lives

swamp: an area with trees that is covered with water at least part of the time

territory: an area of land that an animal considers to be its own

FOR MORE INFORMATION

BOOKS

Emminizer, Theresa. *Deadly Rhinoceroses.* New York, NY: PowerKids Press, 2021.

Scally, Robert. *All About African Rhinos.* Hallandale, FL: EZ Readers, an imprint of Mitchell Lane Publishers, 2019.

WEBSITES

Rhino

www.worldwildlife.org/species/rhino

Read about the dangers rhinos face in the wild and find many pictures and videos here.

Rhinoceros

kids.sandiegozoo.org/animals/rhinoceros

Check out the San Diego Zoo website and learn all about rhinos!

Publisher's note to educators and parents: Our editors have carefully reviewed these websites to ensure that they are suitable for students. Many websites change frequently, however, and we cannot guarantee that a site's future contents will continue to meet our high standards of quality and educational value. Be advised that students should be closely supervised whenever they access the internet.

INDEX